IMAGES
*of America*

# NAVAL STATION
# NORFOLK

IMAGES
*of America*

# NAVAL STATION
# NORFOLK

Hampton Roads
Naval Historical Foundation

ARCADIA
PUBLISHING

Copyright © 2014 by Hampton Roads Naval Historical Foundation
ISBN 978-1-4671-2027-2

Published by Arcadia Publishing
Charleston, South Carolina

Printed in the United States of America

Library of Congress Control Number: 2013933738

For all general information, please contact Arcadia Publishing:
Telephone 843-853-2070
Fax 843-853-0044
E-mail sales@arcadiapublishing.com
For customer service and orders:
Toll-Free 1-888-313-2665

Visit us on the Internet at www.arcadiapublishing.com

*This book is dedicated to every man and woman, military and civilian, who served the United States at Naval Station Norfolk.*

# CONTENTS

# ACKNOWLEDGMENTS

Naval Station Norfolk has many advantages, from its location at the edge of the "world's greatest harbor" to the dedication of thousands of people who live and work there. It has also benefitted from having a museum keep its history. The Navy established the Hampton Roads Naval Museum on the grounds of Naval Station Norfolk in 1979. The commander in chief of the Atlantic Fleet at the time, Adm. Isaac C. Kidd Jr., endorsed the museum by stating that it would "present a propitious occasion to publicize and enhance the Navy's image." The museum's mission to collect, preserve, and interpret the history of the Navy naturally included the history of its home, the Naval Station. The museum staff has welcomed millions of visitors to its exhibits and programs, thus fulfilling Admiral Kidd's hopes.

The museum itself has benefitted from a group of talented historical professionals under the direction of Michael E. Curtin and later Elizabeth A. Poulliot. Under Poulliot's direction, the museum relocated to a downtown maritime center in 1994, achieved American Alliance of Museums accreditation in 2008, and even managed a US Navy battleship, the *Wisconsin* (BB 64), for nearly a decade from 2000 to 2009.

The museum's invaluable supporter has been the Hampton Roads Naval Historical Foundation, a nonprofit foundation incorporated in 1983. The foundation has provided significant financial and material support to the museum over the years. The foundation welcomes members, whose dues help support an active volunteer corps and bring the museum to life.

It is the hope of the foundation that the reader will be inspired by the story of the Naval Station to join us in understanding, celebrating, and preserving the history of the Navy. The reader is encouraged to find out more about the foundation at http://hrnhf.org.

The book before you contains a small sample of the historical materials that the museum has assembled over the decades. Joseph M. Judge, curator of the Hampton Roads Naval Museum, located images for this book. Brian Sagedy from the University of Mary Washington contributed to the editing and layout of this book. All images are from the collections of the Hampton Roads Naval Museum.

# INTRODUCTION

"From Humble Beginning to Largest Naval Complex in the World"—thus a 1967 newspaper summed up Naval Station Norfolk's first 50 years of growth. Decades have intervened since, but the newspaper editor's headline remains a valuable shorthand summary of the growth of the Sewells Point naval complex.

The "Humble Beginning" was a piece of marshy land first inhabited by Native Americans and then by English colonists, who became Americans in the 18th century and fought a civil war in the 19th. Sewells Point saw plenty of human drama over those centuries, especially during the Civil War, but always reverted to a place of quiet agriculture. That changed in 1907 when the Jamestown Exposition, a world's fair celebrating the 300th anniversary of the Jamestown colony, transformed the area. The fair brought months of excitement to southeastern Virginia. While the exposition was not a financial success, it did gain national attention for the region before it closed.

After 1907, Norfolk attorney Theodore Wool undertook the task of finding some positive use for the vacant fairgrounds. He decided to lobby the Navy on the benefits of the exposition site. In a small booklet called *Reasons*, he listed the advantages of Hampton Roads to the Navy: the Chesapeake Bay had deep anchorages and was normally ice-free, there was plenty of vacant land for expansion if needed, the climate in Virginia supported year-round military operations, and rail and maritime transportation networks were in place.

Wool's arguments did not bear fruit until World War I when Pres. Woodrow Wilson endorsed the building of a naval operating base in Hampton Roads. An act of Congress, approved June 15, 1917, authorized the president to commandeer the tract of land that had been the site of the Jamestown Exposition. On July 4, 1917, engineers began staking out boundaries and dredging acres of mud on Sewells Point. On October 12, 1917, a little less than four months from the date of approval of the act of authorization, the first regiment moved to the new Naval Operating Base, with appropriate ceremonies. From this date, the population of the station increased rapidly. New regimental units were being completed and turned over to the commandant every week or two. From a complement of 1,669 on October 17, 1917, the new station (Hampton Roads) had increased to a total of 12,693 by November 27, 1918.

The pressures of war produced explosive growth on the base, seen in barracks buildings, mess halls, lavatory buildings, storehouses, water systems, lighting, roads, and walks. Not to be overlooked were three miles of standard-gauge railroad to afford access to the base, the clearing of approximately 400 acres of ground thickly covered with underbrush, and the development of a system of roads to connect the development with the county road system of Norfolk.

Later, an especially interesting building appeared, known as the "USS Electrician," which was designed as a battleship. Besides classrooms, this school building contained many shipboard electrical devices such as searchlights, signals, cranes, and turret moving and ventilating machines, all operational for training purposes.

Once established, the base became the most important naval installation in Hampton Roads and, by extension, one of the most important in the world. It survived lean years of defense spending in the 1920s and the Depression years of the 1930s until it began another period of explosive growth as the nation prepared for World War II. In line with the development of the area as the major naval base on the Atlantic Coast, the Navy gradually replaced temporary facilities with permanent structures. In 1939, Norfolk was home to the largest of all the Navy's training stations.

The need for even further expansion of the station was not long in appearing. On July 19, 1940, in response to the spread of war across the globe, Congress authorized the "two-ocean Navy." The national effort turned to speed in defense production, and American industry began to produce ships, machines, and equipment at an unprecedented rate. Trained men were needed in ever-increasing numbers, and their flow from the training establishments had to be geared to the flow of ships from the building yards and of equipment from the factories. The supply of personnel became a problem in logistics. Recruiting rates were stepped up, and at Norfolk, 11 new barracks, a drill hall, and attendant auxiliary facilities were begun.

Training sailors for the fleet was not the only wartime concern of the base. The Navy had to satisfy the logistic demands of a Pacific and an Atlantic war, each larger than any previous war. When the Japanese attacked Pearl Harbor, there were only two continental supply depots in commission—one at Norfolk and the other at San Diego. Quickly, the focus of supply construction was on Norfolk, which experienced major expansion that went beyond the limits of the base to the outlying region. In addition to supplying the fleet, Norfolk became, with the passage of the Lend-Lease Act, a supply center for Allied shipping. British trawlers were among the first vessels to use this service. Most of them, built as coal-burning fishing boats, were used for submarine patrol off the East Coast at a time when enemy submarines were sinking many merchant ships in the Atlantic. The same facilities were used by French, Dutch, Canadian, Russian, Greek, Norwegian, Colombian, and other vessels.

The end of the Second World War did not result in the severe defense spending cuts that marked the end of the First World War. The world remained a dangerous place as the Navy confronted the threats arising with the onset of the Cold War. Naval units from Hampton Roads steamed to the front lines of this war, patrolling far-off stations and fighting in conflicts across the globe. These conflicts included wars in Korea and Vietnam and crises in Cuba and the Persian Gulf, among a host of others. The Cold War, a struggle between two economic systems and two different ways of life, ended when the Soviet Union collapsed in 1991.

The Naval Station, as the naval complex was known at the dawn of the 21st century, continues its most important function of preparing the Navy for combat operations in support of US interests around the world. For nearly a century, that mission has been the unchanging character of Sewells Point.

# One

# BEFORE THE NAVY

The Norfolk Naval Station, one of the most important military installations in the world, is located on Sewells Point, a piece of land fronting the historic harbor of Hampton Roads in southeastern Virginia. Hampton Roads, "the world's greatest harbor," is the meeting place of the James, Elizabeth, and Nansemond Rivers and the Chesapeake Bay. Hampton Roads is a condensed version of the name "Southampton's Roadstead," given by the region's settlers to the anchorage. (The Earl of Southampton was an investor in the Jamestown colony. "Roadstead" is an old English word for a protected harbor.)

Southeastern Virginia has long been the site of significant naval activity. The Norfolk Naval Shipyard (originally known as the Gosport Navy Yard) is located on the western bank of the Elizabeth River in the nearby city of Portsmouth. This yard witnessed Navy shipbuilding and repair for most of the 19th century.

Most famously, Hampton Roads was the location of the Civil War battle between the ironclad ships USS *Monitor* and CSS *Virginia* (or *Merrimack*). The battle was the most dramatic but not the only important naval action in the harbor during the war. Hampton Roads was a major staging ground for the Navy and for combined operations against the Confederate coast and Richmond. Early in the war, the Confederates fortified Sewells Point, the first military use of the site.

Sewells Point remained a sleepy place until 1907. That year marked the 300th anniversary of the founding of the Jamestown colony. To celebrate the event, local citizens staged a world's fair called the Jamestown Exposition. It was constructed on Sewells Point and featured exhibits, amusements, military displays, and famous visitors like Booker T. Washington and Mark Twain.

Many politicians and naval officers also visited the fair, which featured a multinational naval review. Most importantly, Pres. Theodore Roosevelt visited twice. He returned to Hampton Roads in December 1907 to launch the around-the-world cruise of a battleship fleet known to history as the "Great White Fleet," which returned to Hampton Roads in February 1909.

**Eighteenth-Century Map of Southeastern Virginia (Detail).** This map shows the wide entrance to the Chesapeake Bay, an excellent maritime highway for early Native Americans and the Europeans who displaced them. Sewells Point, named after early settler Henry Sewell, is marked "Sowels Pt." on this map. Irregular spellings of the place, with and without apostrophes, continued for some time. By the late 20th century, the Navy identified it as Sewells Point.

**Native Americans in Virginia, as Europeans Saw Them.** The first inhabitants of southeastern Virginia were Native Americans. This print, from a 1721 book published in Amsterdam, is one of many based on the 16th-century paintings of the English explorer and artist John White. The Native Americans' presence over centuries in the Virginia Tidewater was quickly erased.

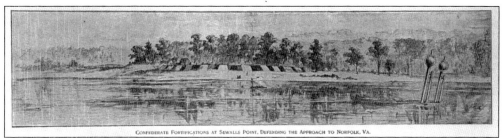

CONFEDERATE FORTIFICATIONS AT SEWALLS POINT, DEFENDING THE APPROACH TO NORFOLK, VA.

**SEWELLS POINT IN THE CIVIL WAR.** In 1861, the Confederates built defensive fortifications on Sewells Point to prevent Union naval forces from descending on the Gosport shipyard in Portsmouth. A Union naval officer described the defenses in 1862: "The works are well constructed, supplied with magazines, many of which contain powder and loaded shells; also bombproof traverses and store rooms."

**BOMBARDMENT OF SEWELLS POINT, MAY 8, 1862.** This 1907 print published in Norfolk depicts CSS *Virginia* (lower left) making her appearance near Craney Island as USS *Monitor* (left center) and other Federal warships withdraw after bombarding Sewells Point. The other US Navy ships presented included *Naugatuck, Dacotah, Seminole,* and *Susquehanna.* Some details, such as the circular fort in the bay, are wildly inaccurate, but the emphasis on Sewells Point as a staging ground for naval activity is not.

**ENGAGEMENT BETWEEN USS MONITOR AND CSS VIRGINIA, MARCH 9, 1862.** This contemporary print of the famous ironclad battle in Hampton Roads identifies Sewells Point on the far left. Neither ironclad survived the year 1862, but the Union established, and maintained, firm control over Hampton Roads.

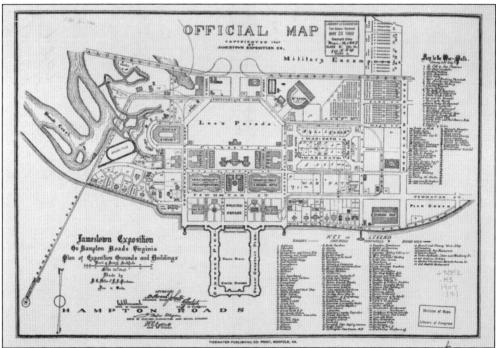

**THE JAMESTOWN EXPOSITION, 1907.** This map shows the extent of development of the Jamestown Exposition, which was not the financial success anticipated by its organizers. The exposition did bring national attention to the region, and fair organizers developed Sewells Point with roads, water, and sewer lines and several large structures. The Navy retained some of the street names, such as Gilbert, Maryland, and Powhatan.

THE PRESIDENT'S PARTY, JAMESTOWN EXPOSITION, JUNE 10, 1907. The carriage at center with two white horses contains Pres. Theodore Roosevelt, who just concluded watching a parade of military units. June 10 was Georgia Day, an important day for the president, whose mother was from the "Peach State." The Georgia state pavilion was a model of Bulloch mansion, her home.

STATE PAVILIONS, JAMESTOWN EXPOSITION. The best-situated structures of the exposition were the state pavilions, which sat immediately on Hampton Roads. These "exploded" models of famous American colonial buildings held exhibits about their respective states. More of them were retained by the Navy than any other building types.

**FLEET REVIEW, APRIL 26, 1907.** On opening day of the exposition, President Roosevelt reviewed assembled warships from several nations. His yacht, the *Mayflower*, is steaming past American battleships with the presidential flag at the top of the mainmast (left center). British armored cruisers are beyond the battleships. Roosevelt's passion and interest in naval affairs was most evident when he assembled a fleet to circumnavigate the world about seven months later—the Great White Fleet.

**A NORFOLK CITIZEN VIEWS THE GREAT WHITE FLEET, DECEMBER 1907.** John Atkinson Steele, a Norfolk railroad worker, took his small boat from Willoughby Spit to view USS *Ohio* (Battleship No. 12) up close as the Great White Fleet prepared to steam out of Hampton Roads.

# *Two*

# BEGINNINGS

After the Jamestown Exposition, a group of local citizens hoped to reap some profit from the dormant site by urging the Navy to purchase the land. This lobbying was fruitless until the United States entered World War I. The conflict called for swift action regarding a naval facility on the East Coast. In 1917, Pres. Woodrow Wilson asked Congress for funding to purchase land on Sewells Point. On June 28, 1917, Wilson issued a proclamation setting aside $1.2 million as payment for the property, with an additional $1.6 million for development of the base. The land purchased comprised 474 acres, 367 acres of which were the old Jamestown Exposition grounds, including several buildings, among them the state pavilions.

Rear Adm. A.C. Dillingham took charge of the construction of the new facility on July 4, 1917. The initial purpose of the base centered on aviation, recruit training, a submarine base, and a supply depot. Dillingham managed thousands of civilian and military workers who cleared the land and then focused their attention on housing, mess halls, and other buildings for naval recruits.

The Supply Depot was initially located in an old resort hotel, the Pine Beach. Two permanent warehouse buildings, one for cold storage, were erected, and the center began service to the fleet. The submarine base had buildings for torpedo and battery storage and a compressor station to supply compressed air to the submarines' ballast tanks.

A major engineering feat was undertaken on the west side of the property, where the flats had to be dredged to allow sufficient depth for ships to berth. The dredged material, eight million cubic yards' worth, was used to create new land, increasing the size of the station to 793 acres.

On October 12, 1917, the Naval Training Station was officially established and thousands of young sailors flooded the new barracks. The name, Naval Operating Base, Hampton Roads, Virginia, was formalized on January 29, 1918. (The abbreviation NOB, although long out of date, continues to be used by locals and sailors.)

**DESOLATION OF SEWELLS POINT, 1917.** By 1917, Sewells Point had reverted to the undeveloped character that it enjoyed for centuries. These images of swamp and abandoned exposition buildings highlight the task facing the Navy as it began work to build an operating base. Thousands of men began work on the installation by simply clearing underbrush and draining hundreds of acres on the new base. All of this effort went on despite two weeks of rain in the initial three weeks of work. The Navy used horse-drawn wagons carrying only half loads to avoid miring down in the mud to remove the debris. These wagons ran in double shifts for a mile and a half to the nearest railroad siding.

**East Wing of the Auditorium, August 1917.** The Navy converted the east wing of the Auditorium, an exhibit structure during the exposition, to an administration building. Its condition was such that workers had to tear the structure down, except for portions of the exterior walls. It is currently designated as Building N-21 in keeping with the Navy's alpha-numeric designations established for the station.

**The Auditorium from the Water, August 1917.** The Navy found roads like this in 1917. The Auditorium, center right, was one of several exposition buildings that the Navy found suitable for government use. It was the focal point of the exposition, and it remained the same for the Navy, becoming the headquarters building. Many other exposition buildings were unusable and quickly torn down.

**ALBERT C. DILLINGHAM AS A LIEUTENANT IN FLORIDA, 1888.** Dillingham took to the business of establishing the base with verve. Recruit Roger Copinger reported to Admiral Dillingham in 1917 and recalled a "fierce-looking officer in a white uniform, with two stars on his shoulder boards" who "burst out of an office, and in one of the loudest voices I have ever heard, demanded to know what we were doing there. . . . Dillingham looked us over very disparagingly, shook his head several times, remarked on the way we wore our hats, our poor shoe shines, our unmilitary bearing, and our general lack of promise; and observed that we should all be at sea learning to sail ships and fire guns."

THE PINE BEACH HOTEL, JULY 1917. Recruit Copinger described the Pine Beach hotel as "a rambling wooden structure in a grove of pine trees. . . . It must have been a fine resort hotel in its time." This remnant of leisure time served the Navy as a supply depot and barracks, but like many of the exposition structures, it eventually gave way to more suitable buildings.

ELEVATED VIEW OF BARRACKS UNDER CONSTRUCTION, AUGUST 1917. Copinger described his first nights in these buildings, which did not have beds at the time: "We were assigned to the newly constructed barracks and the men who did not have hammocks drew them from 'small stores.' Very few had been used to sleeping in hammocks, and the first few nights our sleep was frequently broken by the noise of someone landing on the deck below."

**SEWELLS POINT IN LATE 1918.** Temporary jetties extend from every part of Sewells Point in this image, showing that the work of dredging and filling is well under way. The immediate waterfront features the Pine Beach hotel (left) and huge supply buildings that were the beginning of a century of service to the fleet.

U. S. NAVAL OPERATING BASE. HAMPTON ROADS. VA.

BOAT DRILL AND SEMAPHORE V: TWO POSTCARDS FROM THE ALBERTYPE COMPANY. The Naval Training Station occupied 268 acres of the base and included barracks, mess halls, two indoor drill halls, and two parade grounds. Recruits received instruction in general education, swimming, rigging, and gyro compass use. More specialized instruction allowed the sailors to become machinist mates, hospital corpsmen, storekeepers, cooks, and bakers. Some recruits progressed enough to become Navy petty officers: quartermasters, coxswains, and gunner's mates. Despite the pressures to provide technical training, music was important, and the base musicians and singers gained a wide reputation in those early days. Singing school was held for one hour every evening, and attendance was compulsory. (The Albertype Company of Brooklyn was founded in 1890 as a postcard and view book publishing company. The company went all over the United States to produce postcards, including the Naval Station.)

U. S. NAVAL OPERATING BASE. HAMPTON ROADS. VA.

**THE COLORS: POSTCARD FROM THE ALBERTYPE COMPANY.** *The Bluejacket's Manual* of 1918 advised recruits on their duties during "colors," the traditional honors Navy personnel render to the flag: "At morning 'Colors' the band shall play the 'The Star Spangled Banner.' All officers and men face the ensign and stand at 'Attention' and the guard of the day and the sentries under arms shall come to the position of 'Present' while the national air is being played. At the end of the national air, all officers and men shall salute, ending the ceremony." If there was no band to play "The Star-Spangled Banner," the station bugler sounded a special "colors" call. The ceremony occurred twice a day, at 8:00 a.m. (usually noted as 0800 in military time) and at sunset. Recruits were further reminded that "All officers and men, either from a boat, a gangway, from the shore, or from a ship, shall salute the national ensign."

**SCENES OF BARRACKS LIFE.** Recruit Copinger described the first winter of 1917–1918 at Sewells Point, which was extremely cold: "Shortly after we reported the weather became even colder. The overhead steam pipe froze, and as a result we had no heat in the barracks or hot water in the showers. We slept in our clothes, and our week end liberty became not only a relaxation but a necessity." The severe winter thwarted contractors who struggled to build bulkheads and dredge the waterfront. Ice on the Elizabeth River and the Chesapeake Bay, an unusual sight in Hampton Roads, disrupted shipping and seaplane operations.

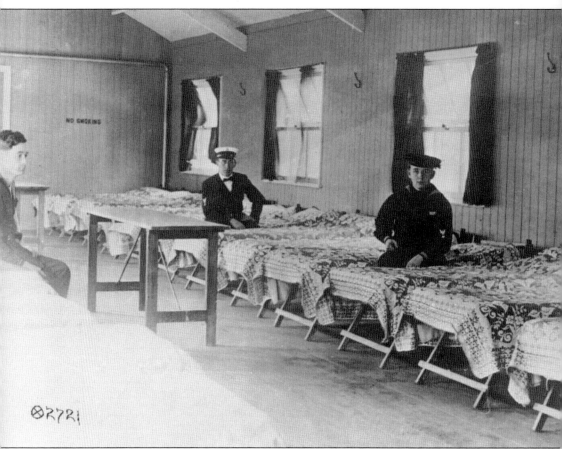

BARRACKS INTERIOR, 1918. Before the advent of steam power and other innovations, the Navy trained recruits by sending them on cruises on square-rigged vessels. Here, the young sailors learned all the basics of seamanship. The technological developments that increased the complexity of ships ended this time-honored system of training. The Navy needed men who could master gunnery, electrical systems, boiler maintenance, and communications. Reese Lukei was one of these recruits. He joined the Navy in Washington, DC, and was assigned to Norfolk. His first night was spent in a barracks like the one seen here: "I was received at the receiving unit (Unit D) and . . . I was assigned a cot to sleep on that night and after seven or eight hours sleep I was awakened by a fellow making an awful racket with a brass horn. It was what they call reveille, which is blown about five o'clock every morning. Well about an hour later the fellow blew the horn again and everybody lined up and marched to the mess hall for breakfast."

**Learning the Art of Knot Tying, 1918.** *The Bluejacket's Manual* of 1918 advised the new recruit that he should be able to tie knots and make splices "with absolute accuracy." It went on to warn, "It can only be learned by practice."

**Platforms for Instruction, 1918.** Recruits climbed these high platforms to practice "heaving the lead line"—Navy terminology for dropping a weight over the side of a ship to determine the depth and character of the bottom. *The Bluejacket's Manual* stressed the importance of the skill: "Every seaman must know how to heave the lead and report soundings correctly."

U. S. NAVAL TRAINING STATION. HAMPTON ROADS, VA.
Clothes Line.

**CLOTHES LINE, ALBERTYPE POSTCARD.** Recruit Lukei described the array of uniforms issued to the man making the transition from civilian to sailor: "A few hours after breakfast I was examined by a couple of doctors and given a lot of clothes including four white and three blue uniforms, a blue hat the shape of a pie plate and three white hats, and a sea bag to keep our clothing in. . . . I was taught to roll all my clothes."

**OFFICER'S MATERIAL SCHOOL, 1918.** The Pennsylvania state exhibit hall, a two-thirds model of Independence Hall, served as an Officer's Material School, a kind of Officer Candidate's School, in 1918. Roger Copinger and other enlisted men embarked on a 90-day course to become naval officers. Copinger remembered the course of study as "seamanship, ordinance, regulations, and navigation," with the day beginning at 5:00 a.m. and extending until lights out at 10:00 p.m. In about three months, Copinger graduated and emerged as an ensign.

26

**SIGNAL SCHOOL PERSONNEL, JUNE 1918.** The commander and his instructors, and their pet cat, pose for the G.L. Hall Optical Co. photographer. Signals instruction included flag signals (semaphore) and International Morse Code signals transmitted by flags, lights, or sounds. (The G.L. Hall Co. produced hundreds of panoramic images like this one in the early years of the station.)

**LOUISE MIRIUM KUSTER, YEOMAN (F), 1919.** During World War I, the Navy began to enlist women as naval personnel for the first time. Nearly 600 yeomen (female) were on duty by the end of April 1917. The Yeomen (F), or "Yeomanettes" as they were popularly known, primarily served in clerical positions, though some were translators, draftsmen, fingerprint experts, ship camouflage designers, and recruiting agents.

27

**THE ELECTRICAL SCHOOL, 1918.** With the world war, the Navy's need for trained men in the operation of modern ships became acute. The Electrical School building housed equipment for training not only in electrical equipment, but also in boilers, batteries, and engines. Rooms were specially ventilated for these purposes.

**STUDYING SHIPBOARD ELECTRICITY, C. 1920.** The trainees in this image are studying a shipboard electrical system during the 30th week of instruction. In the background is a diagram for a master compass. The electrical power used in this classroom was provided by steam generated by boilers in the same building. Other students were at work studying these same boilers.

**FORGE SHOP, C. 1920.** Trainees are in the forge shop during the fifth week of instruction. The sign in the background beside the door bears the title "The Blacksmith's Anvil." Blacksmiths in training were reminded of the seriousness of fire on a ship: "Have a large tarpaulin ready to spread on the deck . . . take care not to allow small pieces of hot metal to fall on the deck and burn holes in it."

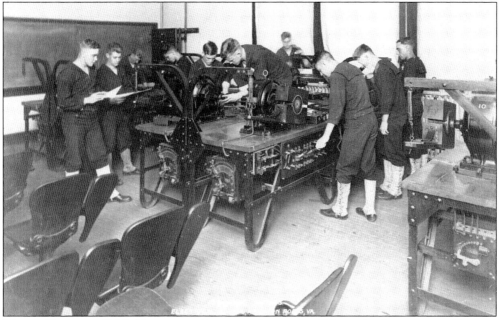

**MOTOR CLASS, C. 1920.** Trainees study motors during the 17th week of instruction. Recruits learned that ships required electricity for air compressors, ammunition hoists, cranes and capstans, winches, pumps, blowers, and radios. Not to be forgotten were its uses for washing machines, dryers, ovens, ice cream freezers, and the all-important electric potato peelers.

**COLORS CEREMONY AT THE COMMISSIONING OF THE ELECTRICIAN, AUGUST 23, 1919.** The USS Electrician was a full-scale mock-up of a battleship, designed to imitate the actual equipment and conditions on a real ship. Searchlights, signals, cranes, turret-moving equipment, and ventilating systems were all operational, allowing apprentice seamen to have thorough training.

**SEMAPHORE DRILL, 1919.** Recruits learned that the two-arm semaphore system was a method in which sailors used two hand flags to spell out messages in code. *The Bluejacket's Manual* warned recruits using semaphore "to face the station or ship squarely and make its call letter. If there is no immediate reply, wave the flags over the head."

**IN RANKS ON THE PARADE GROUND, 1919.** The naval professionalism evident in this image was won at a hard price, according to recruit Lukei: "The next morning we were marched out on a field and just march, marched and marched turning corners and everything. After several days of drilling we were each assigned a rifle, which seemed to weigh a ton, and a bayonet and marched around with the rifles on our shoulders and bayonets hanging from our sides."

**CURTISS N9 AIRPLANE IN FRONT OF THE ADMINISTRATION BUILDING, MARCH 1919.** By 1919, the chaos and debris of the exposition's remnants had given way to the Navy's idea of order. The N9 aircraft in the foreground was the Navy's primary training plane during World War I.

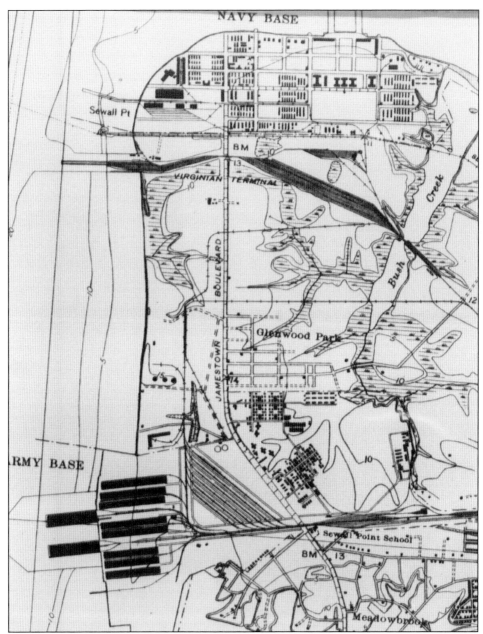

MAP OF SEWELLS POINT IN 1918. The Navy's burgeoning early development of the area can be seen in this image. The waters surrounding Sewells Point were shallow flats, unsuitable for the Navy's ships. At the same time, planners began to feel that the land area available was too small for all the functions to be undertaken at the base. The need for deep water access and increased land space suggested a single solution to the Navy's engineers. Contract workers built bulkheads from the shoreline and dredged sand and mud away from the water side to create deeper water. They then deposited this sand and mud on the land side of the bulkhead, thus expanding the space for building. Suction dredges pumped an unbelievable eight million cubic yards of sand, clay silt, and mud. The depth of water for the submarine base was increased to 25 feet. The depth on the waterfront, where capital ships would soon berth, was increased to 35 feet.

# *Three*

# TAKING FLIGHT

One hundred and fifty acres of the Naval Operating Base was set aside as a landing field for air operations. The sheltered area of the bay was also a welcoming place for seaplane operations.

On October 17, 1917, a naval aviation detachment operating at Newport News, Virginia (across the harbor from Sewells Point), received orders to proceed to the new base. That month, the Navy flew seven seaplanes across the water, accompanied by 5 officers, 18 students, and 20 mechanics. Initially the seaplanes, when not flying, were tied down to stakes in the water. Two weeks later, canvas hangars were in place. Soon the small detachment grew in size. During one week in November, Navy fliers spent over 111 hours flying 157 flights in 26 machines. On November 30, Lt. Comdr. P.N.L. Bellinger took command.

Increases in planes and personnel resulted in new hangars, an administration building, and a medical structure. Aviation personnel also operated a lighter-than-air division, which included dirigibles and balloons primarily used for patrol work. A Construction and Repair Department was the beginning of a tradition of aircraft work that lasted throughout the 20th century. Flying boat service was established between Norfolk and Washington, DC, in November 1917, after the officer in charge made the trip in two hours and 30 minutes.

The Naval Air Station was formally commissioned as subordinate command of the Naval Operating Base on August 27, 1918. Its purposes were training student officers; instructing enlisted personnel in construction, repair, and maintenance of aircraft; conducting patrol flights along the eastern seaboard; and experimental work in seaplane operation, which included the different uses for radios, motors, and stabilizing devices. By 1919, a total of 1,000 men were assigned to NAS Norfolk.

**BULKHEADING AND FILLING AT THE NAVAL AIR STATION, AUGUST 1918.** The rectangular outline of the former Jamestown Exposition lagoon, once surrounded by piers, is now a basin framed by two small peninsulas. A lighter-than-air hangar is at the upper right.

**CHAMBERS FIELD PRIOR TO LANDFILL, AUGUST 1918.** The station's first field for air operations was built up by landfill that displaced the water in the bottom part of this image. In the 1930s, it was named after Washington Irving Chambers, a naval officer who played a major role in the development of naval air power.

**UNIT V SHOWING CONDITION OF THE LANDFILL, SEPTEMBER 1918.** Unit V, the well-defined rectangular area at the top of the image, was the Navy's term for the industrial area of the NAS. Eventually, NAS Norfolk would grow to reclaim all the water in the center of this image.

**ADMINISTRATIVE HEADS OF THE NAVAL AIR STATION, JANUARY 1919.** Front and center is Lt. Comdr. P.N.L. Bellinger. He was the station's first commanding officer. Later, he commanded the Navy's first transatlantic seaplane squadron, and was naval air commander at Pearl Harbor during the Japanese attack in 1941.

**CANVAS HANGARS AT THE EARLY AIR STATION.** Flying during World War I had to serve the dual purpose of patrolling the coast for German submarines and training new pilots. When a U-boat sighting was reported, all training ceased while the instructors flew out to scout the enemy ship.

**WOODEN RAMP FOR SEAPLANES.** The clock tower in the background is the Pennsylvania Building, serving at this time as the Officer's Material School. By the end of the war, NAS Norfolk had trained over 600 officers and thousands of mechanics. It logged over 4,800 patrolling hours, sighted no enemy craft, and lost no aircraft.

**AERIAL VIEW OF NAS NORFOLK, 1919.** The lighter-than-air hangar is at left, with barracks stretching beyond. The lighter-than-air contingent consisted of an assortment of dirigibles, both rigid and nonrigid, and kite balloons, which operated while moored to the ground or a ship and were used in coastal antisubmarine patrols. The dirigible detachment consisted of 32 pilots by the end of the war. These aircraft required extreme care in their operation, as they were filled with highly explosive hydrogen gas. In 1918, one dirigible hit a power wire adjacent to NAS Norfolk and ignited, threatening the entire installation before the resulting fires were contained.

U. S. NAVAL OPERATING BASE. HAMPTON ROADS. VA.

A Curtiss HS-2L Seaplane Leaving for Patrol and Returning to its Hangar, 1918. The HS-2L, seen above, was the Navy's standard single-engine patrol and training seaplane immediately after World War I. In the image below, Navy Patrol Boat SP-484 is operating. Despite the many reported sightings of German submarines, the East Coast was never seriously threatened by them. Many of the ships reported sunk by enemy submarines were actually victims of mines or even natural ocean hazards.

A Seaplane Passes under the Old Jamestown Exposition Arch. In the early days of NAS Norfolk, the arch, a graceful feature of the exposition's waterfront, marked the site of seaplane operations. It was soon torn down.

Hangar for Lighter-Than-Air Operations, 1918. Navy personnel are inflating a nonrigid pressure airship inside the hangar—an aircraft popularly called a blimp. The Navy began ordering blimps early in World War I. The blimps patrolled the coasts and guarded convoys against submarine attacks. The gasbag or envelope was maintained by a combination of lifting gas (which inflated it) and smaller ballonets inside the main envelope.

**NAVY NONRIGID DIRIGIBLE C-3.** The C-3 was one of the C-class airships built in 1918. This huge aircraft, 192 feet long and 50 feet high, caught fire above NAS Norfolk in 1921 due to leaking hydrogen. The crew managed to bring the dirigible to the ground and escape just before it exploded.

**KITE BALLOON, FEBRUARY 1919.** The Navy operated these balloons for observation purposes from 1917 until the 1920s. Almost all of them were directly purchased from Goodyear Tire and Rubber and Goodrich Tire and Rubber.

**A Nonrigid B Series Airship.** The B series of airships represented the Navy's first attempt to design and produce a lighter-than-air craft. The service ordered 16 of them, which were delivered between June 1917 and July 1918. The car underneath this craft was a modified aircraft fuselage with two separate cockpits. It had a water-cooled engine in the nose, which drove a single airscrew. Floatation bags replaced wheels on the bottom of the fuselage. It was suspended from the blimp by a cable—an important fire precaution that separated the engine driving the craft from the gas in the envelope.

**A Curtiss H-12 Aircraft on the Seaplane Ramp, May 1918.** The Curtiss H-series aircraft were patterned after a commercial flying boat designed for transatlantic voyages in 1918. The design had great influence on subsequent flying boats up to World War II.

**A Curtiss H-16 Airplane Sinking, July 1918.** The Navy took a Curtiss design and from it manufactured H-16 aircraft at the new Naval Aircraft Factory in Philadelphia—the first airplanes built there. Some of these remained in service until 1928.

**AN EXPERIMENTAL CURTISS L-TYPE AIRPLANE IN THE BASIN, JULY 1918.** NAS Norfolk tested aircraft and also experimented with radio communications and signaling, new motors, and other devices. Note the Jamestown Exposition arch in the background.

**A TRACTOR PULLING A SEAPLANE, AUGUST 1918.** NAS Norfolk not only trained pilots, it instructed personnel in the construction and maintenance of Navy planes. Part of a pilot's training consisted of assembling an aircraft. By 1918, this training was abandoned with the establishment of an aviation mechanics school.

**MEASURING THE DRAFT OF A CURTISS R-TYPE AIRCRAFT, AUGUST 1918.** The Navy used Curtiss R-type aircraft for scouting, observation, and training. These seaplanes had two cockpits. The pilot sat in the rear while the observer sat in front—somewhat impractical since the observer's vision was restricted by the wings directly above and beneath him. In an experiment, a version of the R-type was modified to carry a torpedo in 1917.

**FORMER JAMESTOWN EXPOSITION STATE PAVILIONS ON THE NAVAL AIR STATION, AUGUST 1917.** These houses, shown in 1917 with supporting gardens, were moved in the 1930s to the other side of the seaplane basin. From right to left, the New Hampshire, Connecticut, and Michigan buildings joined other state pavilions as officers' housing.

# *Four*

# PEACETIME CHANGES

During the 1920s, the base adjusted to peacetime operations. The Navy got smaller, and 1926 marked a low point when only 560 recruits learned to "heave the lead line" at Sewells Point. Despite the reduced nature of operations, Naval Operating Base, Hampton Roads, was considered one of the largest naval facilities in the world. Responsibilities for taking care of Navy ships in the Atlantic rested solely on Norfolk.

The Public Works Department soldiered on under postwar funding limitations. In 1926, workers carried out extensive repairs to Piers 2 and 3. Other projects included repairs on heating lines, electrical feeders, sewer and water systems, telephones, and streets. The waterfront was full of projects like maintenance of dredged channels and basins. Not the least of the problems was a plan to eradicate swarms of mosquitoes that had plagued the base from its foundation.

NAS Norfolk turned its attention to the routine training of pilots. In 1923, engineers tested an arresting device to help pilots training for landing on the Navy's first aircraft carrier, USS *Langley*. Workers also constructed a carrier landing platform on the NAS landing field that pilots used for practice. NAS Norfolk established its own supply department, which operated from a small frame building. Lighter-than-air operations, one of the more interesting sights on Hampton Roads, continued until 1924.

On October 28, 1922, the Navy opened the base to the public in the first Navy Day, entertaining thousands of civilian visitors with a dress parade, aerial and submarine exhibitions, Marine drills, and tours of many buildings. Navy Day became an eagerly awaited feature of life in Norfolk for several years.

**MAIN GATE OF THE NAVAL OPERATING BASE, 1929.** In 1923, the secretary of the Navy ordered a study to determine the capacity of the Navy's bases in war and peace. The fleet was better off than the shore facilities—only Hampton Roads met the Navy's requirements for possible war.

**THE ADMINISTRATION BUILDING, 1929.** Sailors watched movies in this stately holdover from the Jamestown Exposition. Movies were held every night except Monday and Wednesday and were free of charge. Sailors could only attend if they had no guard to stand, no extra duty, or no dirty clothes.

**Pier 2 with USS Langley, 1923.** The *Langley*, the US Navy's first aircraft carrier, was converted from a collier. The work was completed at the Norfolk Navy Yard between 1920 and 1922. The ship carried the nickname "the Covered Wagon."

**MACHINIST MATE WALTER E. PARKER WITH COMPANIONS.** Walter Ernest Parker arrived at the Naval Station in 1922 to attend machinist mate's school. He graduated with honors and supplemented his training by running track and boxing for the Navy.

**HAMPTON ROADS NAVAL TRAINING STATION COLOR GUARD, 1929.** Besides the national colors, the Color Guard carries the Navy Battalion Flag. The Color Guard was front and center at dress parade, including the last parade marking the completion of training. This event was described by Reese Lukei as "the most thrilling of all . . . held Friday afternoon and then on Saturday morning before some of us left for home for a twelve day visit."

**H-12 AIRCRAFT IN PERMANENT HANGARS, EARLY 1920s.** NAS Norfolk began to assume a permanent look after the war. In addition to permanent hangars, the Navy built administrative offices for the station and fleet units.

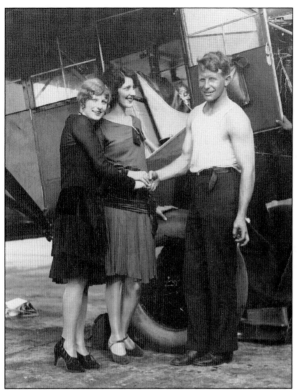

**GOOD LUCK, 1928.** Here, local belles Charlotte Hilliard and Joanne Barnes wish pilot Bernt Balchen good luck. Balchen, a Norwegian, was the pilot on Adm. Richard E. Byrd's Antarctic expedition in 1929. Byrd outfitted the expedition in Norfolk.

**GALLAUDET D-4 AIRCRAFT ON THE SEAPLANE RAMP, OCTOBER 1922.** The Naval Air Station continued to be a center of aviation experiments with planes like the D-4, manufactured by the Gallaudet Company of Norwich, Connecticut. This unusual plane featured the engine amidships, which drove a propeller mounted on a ring encircling the fuselage.

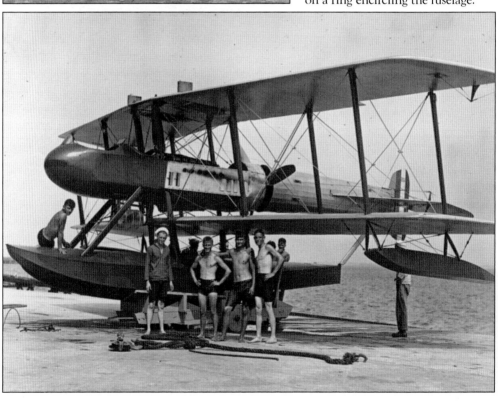

**ASSEMBLY AND REPAIR HANGAR BUILDINGS ON THE NEW LAND EXTENSION, MAY 1923.** The postwar years resulted in an emphasis on aviation production and experimentation, which led to an increase in industrial work. The Navy addressed the need for industrial expansion by filling in ground to the southeast of the landing field where new hangars and industrial shops were located.

**LCDR Harold Terry Bartlett, USN, Flight Commander of the PN-10 Nonstop Flight to Panama.** On November 23, 1926, Bartlett led two Navy PN-10 flying boats out of Hampton Roads in an attempt to make a nonstop flight to the Panama Canal Zone—2,060 miles. Engine trouble forced one down in the Caribbean Sea. The other, after a stop for lubrication trouble, arrived in Panama on November 26.

**Boeing F3B Aircraft at Chambers Field, Late 1920s.** These carrier-based fighter-bombers served the Navy on the early carriers *Langley*, *Lexington*, and *Saratoga*. Experimental detachments from all three of these carriers added to the station's workload at the time.

CAPTAINS STAFFORD H.R. DOYLE (LEFT) AND HARRY E. YARNELL AT THE NAVAL STATION, 1922.
Captain Yarnell was a Spanish-American War and Great White Fleet veteran who pioneered carrier fleet tactics. Captain Doyle won the Navy Cross for transporting troops to Europe in World War I.

**THE ASSEMBLY AND REPAIR DEPARTMENT IN BUILDING V-28, 1927.** At this time, the Assembly and Repair Department employed no civilians. Sailors in this image are working on Vought OS-2 Corsair aircraft. A contemporary Navy report described the function of the department: "This department can disassemble and assemble airplanes and seaplanes and make the necessary alignment of all parts. Hulls, fuselages, floats, wings, struts, etc. can be thoroughly overhauled,

repaired and renewed. . . . Engines and accessories, such as magnetos, batteries, carburetors, can be completely overhauled, repaired and tested. The propeller shop can tip, repair, refinish and balance propellers. The metal shop is equipped to make any kind of sheet metal fittings and repair metal parts of planes . . . The instrument shop can test and make ordinary repairs to all types of aircraft instruments. Radiators can be repaired, tested and cleaned."

## Program

Today is NAVY DAY. The NAVY has set this day apart to be AT HOME TO THE PUBLIC. YOU are invited. The NAVY welcomes you as its GUESTS.

1:00 to 4:00 P.M. School Buildings, Barracks, and Air Station open for inspection.

1:30 to 3:00 P.M. Deep Sea Diving, - Pier No. 7.

1:45 P.M. Lifeboat Drill, - Lagoon. Gun Loading Drill, - Drill Hall.

2:00 to 2:30 P.M. Formation Flying, - Operating Base, Norfolk, and Portsmouth Areas.
Artillery Exhibition and Silent Manual of Arms Drill, - Parade Grounds, Training Station.

2:00 P.M. Football Game, - Naval Training Station vs USS Northampton, - Stadium, Training Station.

2:45 P.M. Between halves of Foobal Game, Exhibition of Stunt Flying.

3:30 P.M. Parachute Jumping.

3:45 P.M. Exhibition Drills and Dress Parade, - Parade Grounds. Training Station.

## Special Points of Interest

Service Schools in Training Station: Rac School, Electrical School, Mechanical Trad

New Barracks "H".

Air Station: Different types of planes on hibit West of No. 1 Hangar with experienc mechanics to explain them. Hangars will open to visitors.

## GENERAL INFORMATION

Please do not exceed the speed lim of 20 miles per hour and obser traffic signs

Park cars where instructed by traffic direct Medical first aid tent East of Administrati Building. In emergencies call Phone 132, O.B. Ladies' rest rooms at Main Gate, Histc Building, Marine Barracks, Administration B ding, Hostess House, and Recreation Buildi at Air Station. Men's rest rooms: Gene School Building, Electrical School, Officer-the-Day, Naval Air Station.

NAVAL OPERATING BASE, NORFOLK, VIRGINIA, NAVY DAY PROGRAM. Navy Days continued to attract thousands of visitors each October. According to a local newspaper, "All roads led to the Navy Base, or so it seemed . . . the great reservation is so immense it was also impossible for the average visitor to get around all of the activities there."

# *Five*

# A DECADE OF SHADOWS

The shadow of the Great Depression covered the Naval Operating Base in the 1930s, just as it covered the rest of the country. The election of Franklin Roosevelt in 1932 brought vigorous government action on many fronts, including the Navy department. In June 1934, the secretary of the Navy established a Shore Station Development Board to manage a program of planned expansion. This board, headed by Adm. Arthur J. Hepburn, reviewed American defense structures during the deteriorating international situation. The Hepburn Board was the basis for the massive shore establishment expansion that took place prior to World War II. In 1938, Congress authorized $4 million for the physical expansion of the Naval Training Station. Civilian employment increased, and the base's payroll reached into the millions of dollars. President Roosevelt visited the base during a tour of naval installations in Hampton Roads in July 1940.

The other shadow in Norfolk during the stressful decade was the coming war, and the base worked busily to prepare for the conflict. In something of a test, the base refueled, restocked, and returned to service 25 ships in one week in April 1939. These 25 ships were but the vanguard of about 100 naval vessels converging on Norfolk. It was commonly asserted that the base's supply center could feed and maintain the entire civilian population of the cities of Portsmouth and Norfolk for a month.

When World War II began in Europe in September 1939, Roosevelt declared a national emergency. Another Roosevelt initiative was the National Emergency Program of September 8, 1939, which resulted in continued growth of all naval activities in the Hampton Roads area.

**Dornier DO-X Seaplane over the Naval Air Station, 1932.** This German flying boat visited the NAS as part of an American marketing tour for international passenger service. Note the hangar roof below, painted "USN Norfolk."

**COMMANDANT REAR ADM. JOSEPH K. TAUSSIG WITH STAFF AND DISTRICT OFFICERS, C. 1938.** Taussig, front row center, led the first group of American destroyers to Europe in World War I. Later, his name graced the main highway connector to the Naval Station.

**HMS NORFOLK AT PIER 2, SEPTEMBER 1933.** When ships like HMS *Norfolk* stopped at Sewells Point, they benefitted from constant work on dredged channels, berths, and basins as well as repairs to piers by the Public Works Department.

**RECRUITS AT THE NAVAL STATION INDOOR RIFLE RANGE, LATE 1930s.** These men are receiving initial sighting instructions on .22 rifles. At the end of the 1930s, approximately 16,000 sailors were shore-based in Hampton Roads—a number set to rise dramatically.

**USS LOS ANGELES (ZR 3) FLIES OVER SHIPS OF THE US FLEET, c. 1930.** Photographed from on board the airship, two of *Los Angeles's* engine cars can be seen in the foreground. Ships below are USS *Patoka* (AO 9), closest to the camera, and the aircraft carriers *Lexington* (CV 2) and *Saratoga* (CV 3).

**BATTLESHIPS AT THE PIERS, JANUARY 1938.** From left to right, USS *Wyoming* (BB 32), USS *Arkansas* (BB 33), and USS *New York* (BB 34) are seen at the piers. All three of these battleships, built well before World War I, survived to serve the country through World War II.

**USS YORKTOWN (CV 5) WITH OTHER SHIPS AT PIER 7, OCTOBER 19, 1937.** The other ships present are, from left to right, USS *Texas* (BB 35), USS *Decatur* (DD 341), USS *Jacob Jones* (DD 130), and USS *Kewaydin* (AT 24).

CHAMBERS FIELD AND SEA WALL, NOVEMBER 1931. In between the world wars, NAS Norfolk served as home base for the aircraft squadrons of the scouting fleet. Two observation squadrons and a torpedo, scouting, and utility squadron of 12 to 18 planes each comprised the permanent force of the scouting fleet.

BOEING AIRCRAFT IN A NAVY DAY FLIGHT DEMONSTRATION, OCTOBER 1933. The annual celebration was still going strong into the 1930s. In 1934, the *Training Station News* placed a large portrait of Pres. Theodore Roosevelt on the cover and proclaimed that "the Navy has set this day apart to be AT HOME THE PUBLIC. YOU are invited."

**An Air Ambulance Returns from Cape Hatteras, North Carolina.** The Naval Air Station actively operated about 30 planes for both land and sea. One large twin-motored seaplane was always kept on ready status for emergency flights.

**Rear Adm. Guy Burrage and Family at the Virginia House, c. 1930.** Admiral Burrage, commander of the Fifth Naval District, was one of the many flag officers to reside in the former Jamestown Exposition pavilions.

**ATLANTIC FLEET DESTROYERS AT NORFOLK, 1935.** From left to right are USS *Schenck* (DD 159), USS *Leary* (DD 158), USS *Dickerson* (DD 157), and USS *Herbert* (DD 160). Civilian visitors walk along the waterfront.

**USS Yorktown (CV 5), September 30, 1937.** The crew of the *Yorktown* parade on the flight deck during the ship's commissioning ceremonies in September 1937. The Marine detachment is at lower left center. Note aircraft tie-down strips laid at regular intervals among the flight deck planks. The *Yorktown*, built at nearby Newport News, sank after the Battle of Midway on June 7, 1942.

**FLEET PLANES LINE CHAMBERS FIELD, JULY 1939.** Willoughby Bay stretches beyond the field to the upper right. Admiral's Row, the Jamestown Exposition flag housing, lines the street on the left. At the bottom of the picture sits the Auditorium and wings, the headquarters complex.

**THE NAVAL OPERATING BASE ON THE EVE OF WORLD WAR II, AUGUST 1939.** The base's expansion into the harbor is evident from this image. Willoughby Spit curls around to the right.

## Six

# WORLD WAR II ON THE NAVAL STATION

When war broke out in Europe in September 1939, the enlisted strength of the United States Navy was about 110,000 men; six years later in 1945, personnel numbered over three million. This expansion had an impact on the Norfolk base where the Navy began work on 11 new barracks, a drill hall, a new auditorium, and other facilities.

The Supply Depot grew to its limits, adding modern piers and a six-story, 1.8-million-square-foot warehouse, the largest government building south of the Pentagon. Schools for electronics, firefighting, torpedoes, welding, and compass work trained sailors in the complexities of modern warfare.

With so many recruits, there was a housing shortage. Rear Adm. Joseph K. Taussig convinced Adm. Ben Morrell, the head of the Bureau of Yards and Docks, to build the Navy's own housing, which was named Benmorrell. Rents were based on rank: a petty officer third class could rent a three-bedroom apartment for $15 a month.

The Navy added new athletic facilities to the base, including baseball fields where stars like Bob Feller and Pee Wee Reese played for Navy baseball teams as they served their country.

Early in the war, Allied planners decided to secure French North Africa—the greatest amphibious operation up to that time. Operation Torch, as the cross-ocean assault was known, was partially planned and launched from Norfolk. Planning took place in the Nansemond Hotel in Ocean View. Landing exercises were conducted in secret at Solomons Island, Chesapeake Bay. Task Force 34, under the command of Rear Adm. H. Kent Hewitt, embarked with the Western Task Force of Gen. George S. Patton Jr. (35,000 troops) from Norfolk to the Atlantic coast of French Morocco. Hampton Roads was also the point of origin for many convoys, and convoy support and escort remained a critically important function. In 1943, the Navy built new piers for these convoy escorts.

The war and its victims also came closer to home. In the spring of 1942, a small number of German U-boats wreaked havoc on East Coast merchant shipping, including offshore Virginia and North Carolina. In June 1942, U-701 mined the waters near the entrance of the Chesapeake Bay. A stepped-up antisubmarine warfare program included patrol flights from NAS Norfolk and subordinate air commands. U-boats sunk off Virginia and North Carolina, naval intelligence breakthroughs, coastal convoys, and powerful antisubmarine weapons ended the submarine threat.

**THE END OF THE AUDITORIUM.** This survivor of the Jamestown Exposition burned on January 26, 1941. It was quickly replaced with a new headquarters building, designated N-26. The new headquarters building allowed the Navy to designate more space for a fleet post office.

**PIERS 1 AND 2, NOVEMBER 1943.** Pier space was a pressing need at the naval base during the war. The Navy needed to alleviate the serious waterfront facilities shortage and provide space for Destroyer Escort Training. In March 1943, the chief of Naval Operations requested the Bureau of Yards and Docks to take emergency action to acquire the necessary land.

**PIERS, APRIL 1940.** In August 1943, the commandant proposed the expansion of three convoy escort piers, each 750 feet in length. The actual construction of the piers was started in late September 1943. The first pier to be completed was in operation by the middle of June 1944, and by the end of July, all three piers were berthing ships.

SECRETARY OF THE NAVY FRANK KNOX PAYS A VISIT TO REAR ADM. MANLEY H. SIMONS, AUGUST 1941. Admiral Simmons was the commandant of the Naval Operating Base and the commandant of the Fifth Naval District. In addition, Admiral Simmons commanded the Chesapeake Task Group of the Eastern Sea Frontier, where he worked to stem the assaults of German U-boats.

THE NEW NAVAL HOSPITAL, JUNE 1942, AT 800 FEET ALTITUDE. These buildings were constructed to be the US Naval Hospital, Norfolk, Virginia. This complex, designated as the NH Unit in the South Annex, would become Atlantic Fleet headquarters in 1948.

**SS San Jacinto Survivors Come Ashore at Norfolk, April 1942.** German submarine U-201 attacked and sank the American passenger ship *San Jacinto* on April 21, 1942. USS *Rowan* (DD 405) rescued survivors of the attack and brought them to the Naval Station. *Rowan* herself was sunk in the Mediterranean in 1943 with great loss of life.

**A Crowded Pier 5, 1944.** Author Fletcher Pratt described the base as "a place of improvisations, impermanence and constant flux, as disorderly as war itself, whose only constant is keeping the ships at sea."

**DAMAGE CONTROL CLASS, OCTOBER 1944.** These sailors came from different parts of the world: two are from Iceland, four from New York, four from Rhode Island, two from Washington, DC, and two, including the chief at the bottom right, from Norfolk.

**LOADING TIRES AT THE PIERS, JANUARY 1944.** World War II focused attention on logistics to a greater extent than any previous war. Part of this focus was due to the increased intricacy of modern ships and planes and weapons, and part of it was due to the great quantities of every item needed to service the enormous number of men, ships, and bases of the wartime Navy.

**GRUMMAN TBF/TBM AVENGER AIRCRAFT PROCEED ALONG WATERFRONT, MARCH 1943.** These planes were hoisted aboard USS *Essex* (CV 9) in preparation for her voyage to the Pacific, where she would win 13 battle stars.

**WOMEN PRESS REPRESENTATIVES ABOARD USS WYOMING (AG 17), OCTOBER 1941.** The press and radio women pose with *Wyoming's* commanding officer, Capt. Van Leer Kirkman, atop her forward 12-inch gun turret while visiting the ship during a press tour of the Naval Operating Base.

**TROOPS BOARDING A SHIP, 1943.** Although U-boats remained a menace throughout the war, by the middle of 1943, the Allies had largely controlled them through technological advances, code breaking, and the production of new ships and aircraft. The Allies staged massive amphibious assaults that assured the defeat of Nazi Germany, including assaults in North Africa in November 1942, Italy in September 1943, and Normandy, France, in June 1944.

**LOADING PLANES ON USS INDEPENDENCE (CV 22), APRIL 1943.** In 1943, *Independence* was the first of a new class of carriers converted from cruiser hulls. She left Norfolk after loading these planes for the Pacific war, where she received eight battle stars.

**SHIPMASTERS AT A PRE-SAILING CONFERENCE FOR CONVOY UGS-13, JULY 1943.** The port officer speaking to the shipmasters is standing in front of a chart of convoy ship positions. The masters in the audience are seated in the same relative positions as their ships' assignments. Allied convoys were identified with letters and numbers—the UGS series sailed from Hampton Roads to Port Said, Egypt.

**CONVOY SHIPMASTERS RETURN TO SHIPS, JULY 1943.** Convoy UGS-13 sailed the next day. It consisted of 97 merchant ships and 29 escort vessels. The UGS series were designated as "fast convoys." UGS-13 departed Hampton Roads on July 27 and arrived in Egypt on August 24, 1943.

**CONVOY UGS-13 IN HAMPTON ROADS THE DAY BEFORE SAILING.** The UGS series was instigated in November 1942 to support Operation Torch but continued until May 1945. The Allied convoy system benefitted from the success of the intelligence war against the Germans, most often by simply being routed away from places where U-boats were known to be operating.

**GUNNER'S MATES FROM USS MASON, JANUARY 1944.** Here, gunner's mates from *Mason* assemble and study a 20-mm machine gun. The Navy commissioned USS *Mason* (DE 529) in March 1944 with a largely African American enlisted complement. *Mason* was employed on convoy escort duties in the Atlantic and Mediterranean through the remainder of World War II. The instructor is Chief Gunner's Mate Rex Ashley. Trainees are, from left to right, Gunner's Mate 3rd Class Albert A. Davis, Gunner's Mate 2nd Class Frank Wood, and Gunner's Mate 2nd Class Warren Vincent.

**FIRST WOMAN ORDNANCE TRAINEE EMPLOYED BY NAVAL SUPPLY DEPOT, JUNE 1942.** Myrtle C. Freeman was part of the base's Supply Depot that expanded from 10 warehouses and 2 piers to 41 warehouses and 13 piers during the war. Its supply functions also extended to outlying regions, such as Yorktown and Williamsburg.

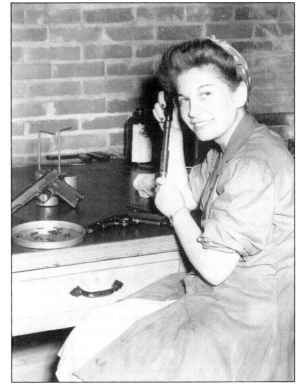

**PRESENTATION OF CITATIONS.** Adm. Royal E. Ingersoll, commander in chief of the Atlantic Fleet, presents Presidential Unit Citations to the officers and men of three destroyers in May 1943. The destroyers, USS *Dallas* (DD 199), USS *Cole* (DD 155), and USS *Bernadou* (DD 153), received the award for their actions in the invasion of North Africa, Operation Torch. Ships in the background are USS *Dyson* (DD 572) and USS *Alcor* (AC 10).

**MCCLURE FIELD, SEEN FROM CENTER FIELD.** During World War II, the station's baseball stadium, opened in 1920, was home to many major league players who had enlisted in the Navy and reported to Norfolk for training. While learning how to be sailors, stars like Bob Feller, Dom DiMaggio, and Pee Wee Reese played baseball for the troops and raised millions of dollars in war bonds.

**THE BAND FROM CARRIER SERVICE UNIT 21 (CASU 21).** The Navy worked hard to provide an elaborate program of recreational activities during the war years; dances, games, movies, exercise classes, and sporting events were all available.

**4**

UNITED STATES OF AMERICA
OFFICE OF PRICE ADMINISTRATION

# WAR RATION BOOK FOUR

Issued to _Charles J. Devine_
(Print first, middle, and last names)

Complete address _Argyle Ave, Lochaven,_

_Norfolk, Va_

## READ BEFORE SIGNING

In accepting this book, I recognize that it remains the property of the United States Government. I will use it only in the manner and for the purposes authorized by the Office of Price Administration.

*Void if Altered* _____
(Signature)

*It is a criminal offense to violate rationing regulations.*

OPA Form R-145

16—35570-1

The stamp reads vertically: VALIDATED FOR COMMISSARY STORE, N.O.B.

WAR RATION BOOK FOUR, OFFICE OF PRICE ADMINISTRATION, 1943. The Office of Price Administration rationed supplies and set prices during the war. Consumers used stamps from ration books like this one to purchase scarce commodities. The stamp on the front reads, "Validated for Commissary Store, N.O.B."

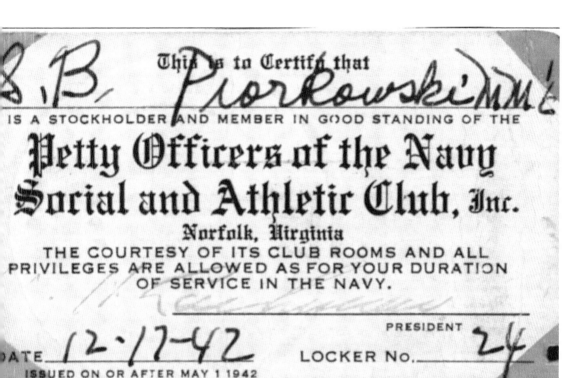

**MEMBERSHIP CARD, 1942.** A number of social clubs sprang into existence during the war designed to provide wholesome entertainment for the thousands of military personnel passing through the region. This membership card was for petty officers of the Navy Social and Athletic Club Inc.

**STAFF MEMBERS OF NAVY OFFICER'S QUARTERS AND CLUBS.** This photograph was taken in the Pennsylvania Building, the former Officer's Material School, which by World War II had become the Naval Station Officer's Club. These men ran Bachelor Officer's Quarters and Clubs throughout the Hampton Roads region.

# Seven

# WORLD WAR II AND THE NAVAL AIR STATION

The Naval Air Station in Norfolk began a period of tremendous growth as early as 1939, stemming from the actions of the Hepburn Board (see chapter five). As a result, between 1940 and 1942, the Navy added 1,441 acres to NAS Norfolk at a cost of $966,000. Part of the development was 352 acres of reclaimed marsh land, built up using silt dredged for seaplane operations in Willoughby Bay. A subsequent contract produced a burst of activity in which workers built landing fields, hangars, storehouses, magazines, barracks, and docking facilities.

Growth was intended to accommodate four aircraft carrier groups, six to ten patrol squadrons, and facilities for compete aircraft overhaul in line with the nation's growing construction schedule.

The staff of the NAS also prepared for the global nature of the coming conflict. As early as fall 1941, British pilots, radio operators, and gunners from HMS *Illustrious* and HMS *Formidable* were training at Norfolk; they would be followed throughout the course of the war by pilots from France, Russia, and several South American countries.

The amount of air traffic controlled gives a good feel for the hectic pace of operations at NAS Norfolk. The flight operations department in 1943 recorded an average of 21,073 flights per month, or about 700 a day. The fighter director school taught fleet communications and tactics, radar operations, and the direction of aircraft from ships.

The worst disaster on the base occurred during the war on September 17, 1943, when 24 depth charges exploded while being moved. Thirty people died and hundreds were injured in the accident. Fifteen buildings on NAS Norfolk were demolished.

The Assembly and Repair Department continued to repair aircraft and overhaul engines as well as provide items such as searchlights, radios, catapult fittings, and oxygen breathing equipment for numerous naval installations. After the country entered the war, the workforce numbered 1,600 military and 3,500 civilians, including many women.

**NAVAL AIR STATION OPERATIONS BUILDING AND CONTROL TOWER, SEPTEMBER 1941.** At the onset of the war in Europe, the NAS Norfolk was operating with two small areas, Chambers Field and the West Landing Field. Expansion began to the east.

**THE STEEL SKELETON OF SEAPLANE HANGAR NO. 2 (SP-2) UNDER CONSTRUCTION, DECEMBER 1940.** Hangars, a new dispensary, runways, magazine areas, and barracks were patterned after existing airfields. The plans for expansion were approved by Capt. P.N.L. Bellinger, returning as commanding officer after 20 years.

INTERIOR OF SP-2, APRIL 1941. In 1941, the construction of new facilities was rushed forward as US involvement in the war grew more likely. These new requirements greatly expanded the scope of construction.

DOUGLAS TBD-1 AIRCRAFT. Douglas TBD-1 aircraft of Torpedo Squadron Five (VT-5) are parked at NAS Norfolk in September 1941. Douglas SBD-3 planes of Bombing Squadron Five (VB-5) are in the background. Both squadrons were assigned to USS Yorktown (CV 5).

**A Grumman TBF Avenger, September 1942.** In the center background sits a Douglas SBD Dauntless. As far as direct action against the enemy was concerned, NAS Norfolk's contribution was in the area of antisubmarine patrols.

**LP 20 under Construction, April 1942.** The Navy wanted to ensure that as many of the new structures as possible were permanent—many of the World War I buildings were "temporary" hangars and workshops, unsafe and costly to maintain.

**ADVANCED BASE "A" TRAINING UNIT, HANGAR NO. 2, MAY 1942.** The hangar combined working and sleeping quarters for 11,000 men. A typical wartime work schedule at NAS Norfolk was two 10-hour shifts per day, seven days a week. Women joined the workforce in increasing numbers to combat acute labor shortages. An apprentice school opened in 1942 to provide training in nine different trades.

**SIGNALS TO A LANDING PLANE, 1944.** Specialist 2nd Class Jane Rockman, a US Navy WAVE, uses a control tower signal lamp to flash landing instructions to the incoming plane. Women returned to general Navy service in early August 1942 after a lengthy effort to establish the WAVES, or "Women Accepted for Volunteer Emergency Service."

**A GRUMMAN F4F WILDCAT ON THE LANDING FIELD, FEBRUARY 1942.** The F4F was one of the outstanding Navy fighter airplanes of World War II. The initial delivery of the F4F aircraft was made to VF-4 (USS *Ranger*) and VF-7 (USS *Wasp*) at NAS Norfolk in 1940.

**SAILORS ON THE SEAPLANE RAMP, JULY 1942.** The seaplane operating area on Willoughby Bay was a major addition at the time; two large hangars, ramps, and barracks were built.

**A MARTIN PBM3 MARINER, NAS NORFOLK WATERS, 1942.** The Mariner flying boat gave the Navy valuable service as a patrol aircraft. A prototype of this airplane, designated the XPMB-1, was sent to the Aircraft Armament Unit at NAS Norfolk in 1941 for armament experiments.

A GRUMMAN F4F WILDCAT AND TBF AVENGER, SEPTEMBER 1942. One of the wartime challenges for NAS Norfolk was to join newly created training units to a number of existing units in order to meet the demand for combat-ready aircrews and maintenance personnel.

INTERIOR OF A MARTIN PBM3 MARINER, SEPTEMBER 1942. This view shows the radio compartment and the cockpit while the Mariner was in flight over Norfolk. It carried a crew of nine.

**CATASTROPHE AT NAS NORFOLK.** At 11:01 a.m. on September 17, 1943, a depth charge slipped off a trailer and the resulting friction exploded that bomb and 23 others. The charges exploded singly or in groups for several minutes. The blast shattered windows up to seven miles away and was heard in Suffolk, 20 miles away. Tragically, the accident killed 30 and injured over 400. One of the dead was Seaman 2nd Class Elizabeth Korensky, the first WAVE to die in the line of duty during the war. Fifteen buildings were destroyed or so badly damaged that they had to be demolished, and 33 aircraft were total losses.

A CONSOLIDATED PB4Y
LIBERATOR READIES FOR
TAKEOFF, DECEMBER 1943.
The Liberator, the Navy's
version of the B-24 bomber,
played a critical role in ending
the U-boat threat during
the battle of the Atlantic.

"OLD" CHAMBERS FIELD
WITH A MYRIAD OF AIRCRAFT,
MAY 1944. Chambers Field
is pictured here at 4,000 feet
altitude. By 1943, the old NAS
Norfolk had become the hub
for a series of outlying airfields,
including Chincoteague,
Oceana, Pungo, and Fentress
in Virginia and several
more in North Carolina.

**Junior Bachelor Officers' Quarters under Construction, July 1942.** In 1943, the Navy created the position of commander, Air Force Atlantic Fleet to provide administrative, material, and logistical support for the Navy's aviation units. This photograph was taken at 500 feet altitude.

**A Confident Navy Pilot.** Lt. Robert L. Brown is seen here posing with the squadron mascot "Scrappy" in the cockpit of his F6F Hellcat.

**A New Hangar under Construction, August 1944.** This hangar, V-88, was photographed at 200 feet altitude. The station had to prepare to operate five aircraft carrier air groups, seven to nine patrol squadrons, the Fighter Director School, and the Atlantic Fleet Operational Training Program for 200 pilots.

**NAS Norfolk Control Tower, 1944–1945.** The view from the inside shows that the tower overlooked a seaplane ramp. Planes on the ramp are Vought OS2U Kingfisher floatplanes. One of those working in the tower is a WAVE.

# *Eight*

# WITHOUT US, THEY DON'T FLY

Naval Air Station Norfolk featured a dedicated aircraft maintenance detachment from the earliest World War I period of activity, when the Construction and Repair Department was part of the commissioned station. In 1922, it was renamed the Assembly and Repair Department and, under that name, met the challenges of World War II.

After the war, technology—and therefore flight speed and the complexity of aircraft—increased dramatically. In May 1948, the first jet squadron qualified for fleet operation. The jets made an immediate impact; from 1947 to 1956, the speed of the Navy's top aircraft increased 165 miles per hour per year. The Assembly and Repair Department, which had not assembled aircraft for some time, was renamed Overhaul and Repair, commonly abbreviated to O&R.

In the early 1950s, the department responded to the requirements of the Korean War, and peak employment reached 7,200 people. It overhauled 200 aircraft each month. Later in the decade, the department embarked on an Engineering Performance Standards program that simplified and standardized methods, materials, tools, and equipment. This program also established standard times for the overhaul and repair of aircraft.

The series of photographs in this chapter are part of a Navy photographic survey of the O&R Department's various shops, compiled in March 1950. During this time, the department comprised 200 buildings (within which were 340 shop and office areas) on 214 acres. Tremendous changes were under way during this period—guns were replaced with guided missiles, and swept wings and afterburner jet engines appeared. The civilian employees, however, were still paid in cash.

**DISASSEMBLY OF AIRCRAFT.**
Operations within the
Disassembly Shop consisted
of disassembling large aircraft,
segregating and identifying parts
and assemblies, and preparing and
attaching engine check-off sheets.

**CLEANING OF AIRCRAFT.**
Operations within the
Cleaning Shop consisted of
painting, stripping, and steam
cleaning aircraft fuselages
and component parts.

**AIRCRAFT FINAL ASSEMBLY.** Operations within the Final Assembly Shops consisted of the progressive assembly of large planes by stages. Some of the operations included installing fuel systems and landing gear, checking hydraulic equipment and systems, and installing electrical equipment and instruments.

**LARGE SURFACE REPAIR.** This image shows the installation of surfaces and control mechanisms in a wing. The Surface Repair Shop also repaired aircraft ribs, spars, fittings and rivets, and applied corrosion inhibitors.

FINAL PAINT. The Final Paint Shop applied, under strictly controlled atmospheric conditions, primary and final finish to the exterior and interior of hulls, wings, and surfaces of planes. The shop also applied insignia, markings, and decals to the aircraft.

ORDNANCE SHOP. Operations within the Ordnance Shop consisted of disassembling, cleaning, examining, repairing, overhauling, modifying, adjusting, testing, and preserving aircraft gun turrets and gear boxes. The shop also worked on aircraft machine guns and rocket launchers.

**METAL SHOP.** The Metal Shop manufactured steel or aluminum parts used in the overhauling, modifying, or repairing of aircraft parts like engine mounts, cowling, and arresting hooks. Workers also conducted welding repairs in this area.

**SALVAGE OPERATIONS.** Salvage operations within this area consisted of disassembling, examining, and identifying parts of aircraft in a condition otherwise unsuitable for use. Materials were reclaimed for use by the various department shops. The shop also maintained records of all this work.

**AIRCRAFT PRESERVATION.** Operations within this shop consisted of preparing aircraft for preservation storage in containers by removing wheels, installing desiccant, and taking readings of atmospheric conditions such as temperature and humidity. The Navy considered this type of storage to be long-term.

# Nine

# INTO A COLD WAR

At the end of the war, the Navy reorganized for peacetime operations. The Naval Operating Base took a new name on November 30, 1945, when it was re-designated as US Naval Station Norfolk. A new command named US Naval Base Norfolk was established on the same date, with US Naval Station Norfolk listed as a separate component. The base commander remained the Navy's chief shore representative in Hampton Roads.

NAS Norfolk continued to operate at near peak levels. It served as the operational headquarters for the Fleet Air Command and, with the emergence of Oceana as a master jet airfield in the 1950s, formed the nucleus of the biggest air base on the East Coast.

Change also came to the Naval Station side of Sewells Point. The Atlantic Fleet came ashore by moving its headquarters from USS *Pocono* to the buildings of the former US Navy Hospital, Norfolk, where it has remained. In 1948, a Public Works Center was established, charged with maintenance of quarters and other buildings and undeveloped areas. The Supply Depot continued to serve the ships of the Atlantic Fleet, and they deployed around the world, including during the wars in Korea and Vietnam.

In the early 1950s, the North Atlantic Treaty Organization (NATO) decided to establish a new Allied Command, Norfolk, Virginia. This command, primarily charged with allied defense of the North Atlantic, was assumed by the commander in chief of the Atlantic Fleet from 1952 through 1985.

The Naval Station, while it remained a closed environment, could not be isolated from the forces that were changing American society. The Navy officially integrated in 1949, but as in the rest of society, meaningful change did not occur at once. In the mid-1960s, Rear Adm. Reginold Hogle ordered an end to discrimination, and established a biracial committee of leading civilians and military officers to address the problem.

**NAVAL BASE PIERS, OCTOBER 1948.** This photograph of 52 ships and harbor craft was taken from an altitude of 2,500 feet. The ships that are present, among others, include LSTs, the battleship *Missouri*, the cruisers *Juneau* and *Mississippi*, destroyers, and destroyer escorts.

**A CEREMONY ON BOARD SUBMARINES, LATE 1950s.** The submarines are moored alongside USS *Orion* (AS 18). From the rear, they are USS *Sailfish* (SS 572), USS *Cutlass* (SS 478), USS *Runner* (SS 476), USS *Cobbler* (SS 344), USS *Argonaut* (SS 475), an unidentified sub, USS *Requin* (SSR 481), and USS *Barbero* (SSG 317).

**A Busy Waterfront, August 1951.** The crowd here is celebrating the arrival ceremony of Destroyer Division 162. After World War II, the Naval Base established the Port Director's office and charged it with maintenance of all piers.

**Gate 2, 1950s.** In 1951, the commander of the base recognized that the Sewells Point area needed a more centralized control of development. A Station Development Board undertook the task of reviewing all projects that might affect Sewells Point and prepared a long-range development plan.

Admit wearer to section reserved for the

# PRESS

On Board USS NEW JERSEY (BB-62)
Pier 7, U. S. Naval Base
Norfolk, Virginia

9:30 a.m. (EST), Monday
**APRIL 12, 1954**

# CHANGE OF COMMAND

UNITED STATES ATLANTIC FLEET

**Representing:**

_____
(Name of Organization)

IMPORTANT: Keep this tag in sight while
on the Naval Base. Press representatives
must be on board no later than 8:30 a.m.,
April 12. Microphones cannot be installed
after that time.

**PRESS PASS, USS NEW JERSEY CHANGE OF COMMAND, APRIL 1954.** *New Jersey,* one year removed from the Korean War, was sharpening her skills with exercises and training maneuvers along the Atlantic coast and in the Caribbean.

**USS MISSOURI (BB 63) AT THE NAVAL STATION, 1950s.** On January 17, 1950, the famous battleship went hard aground on Thimble Shoals in Hampton Roads. Her momentum forced her well into the shallows, raising her hull several feet above its normal waterline. Freeing the ship involved offloading ammunition, supplies, and fuel and the concentrated efforts of many tugs, dredges, and divers. *Missouri* finally returned to the channel on February 1. The ship left a permanent depression in the watery bottom, known as the "Mo Hole," where locals still seek good fishing.

**USS *Forrestal* (CVA 59) and USS *Independence* (CVA 62) at Pier 12, July 1959.** At this time, *Forrestal* had returned from her first deployment to the Mediterranean, while the newly commissioned *Independence* prepared for training in the Atlantic.

**MARTIN MARLINS IN FLIGHT, 1950s.** These planes are from Squadron VP 56. They are flying over Willoughby Spit, a narrow strip of land adjacent to the Sewells Point complex. The Martin P5M-1 Marlins seen here were the Navy's last operational flying boats.

**THE P2V-1 TRUCULENT TURTLE EXHIBITED AT GATE 4 OF NAS NORFOLK.** The "Turtle" set the world's long distance nonstop flight record in 1946 when it flew from Perth, Australia, to Columbus, Ohio (11,236 miles in 55 hours and 18 minutes). The Turtle was a prominent feature at NAS Norfolk from 1953 to 1977, when it moved to the Naval Aviation Museum in Pensacola, Florida.

**NAS Norfolk Float, Armed Forces Day Parade, May 1961.** The tradition of Navy Days earlier in the century changed over time to Navy participation in community events and special occasions in the neighboring cities, such as parades like this one. The most notable event during this period was Norfolk's Azalea Festival, which honored the NATO alliance.

**Foreign Object Damage (FOD) Clean Up, April 1962.** One hundred fifteen men from the Aircraft Maintenance Detachment followed this procedure every Thursday at 12:30 p.m. They are looking for any object that could potentially injure personnel or damage aircraft. FOD management remains a concern for all air operations today.

**BARGE RECEIVING STATION, NAS NORFOLK.** This barge crew hooks up the shore side pipeline to discharge its cargo of jet fuel. Carefully observing are the fuel branch supervisor and the fuel inspector (standing at center).

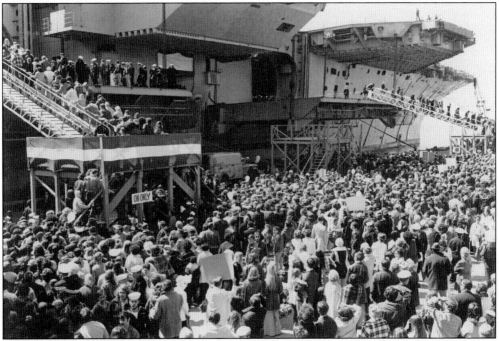

**USS AMERICA (CV 66), MARCH 24, 1973.** Thousands welcome USS *America* back from a 10-month tour in southeast Asia. This tour was the sturdy carrier's sixth major deployment since her commissioning in 1965.

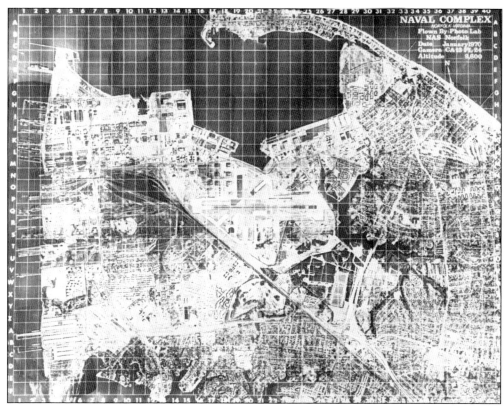

NAVAL COMPLEX
Flown By: Photo Lab
NAS Norfolk
Date: January 1970
Camera: CA 13 FL 24
Altitude: 9,600

**THE NAVAL COMPLEX IN 1970 FROM 9,600 FEET ALTITUDE.** During the 1960s and early 1970s, the Naval Station supported many ships that were deeply involved in the conflict between Communist North Vietnam and South Vietnam.

**CRANES FROM THE PUBLIC WORKS CENTER AWAIT DUTY ON THE WATERFRONT, 1970s.** The base newspaper, the *Seabag*, informed readers that the Public Works Center was constantly at work repairing over 1,000 buildings, 85 miles of roadways, 10 piers, 4 railroad locomotives, and 6 floating derricks.

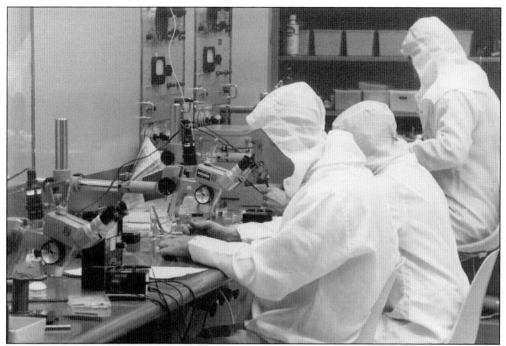

**TECHNICIANS REPAIR EQUIPMENT IN A CLEAN ROOM AT THE NAVAL AIR REWORK FACILITY (NARF).** In 1967, the venerable O&R Department of NAS Norfolk was renamed the Naval Air Rework Facility. Later, it was renamed yet again as the Naval Aviation Depot.

**AIRCRAFT CROWD INTO HANGAR LP 2, SEPTEMBER 1967.** These aircraft were seeking to avoid the damaging winds of Hurricane Doria. Then, as now, hurricane preparedness was an important task for the Naval Station.

**Painting USS** *Inchon* **(LPH 12).** Seaman Henard Gothard gives a fresh coat of paint to the
22,500-pound anchor attached to USS *Inchon* (LPH 12). The ship was preparing for a NATO
mine countermeasure exercise in the North Atlantic.

**FIGHTING A SIMULATED FIRE, NAS NORFOLK.** Exercises like this one helped NAS personnel be ready for any possible emergency. In the late 1960s, the tempo of operations at NAS Norfolk averaged 169,000 operations per year, including 18 million pounds of air cargo and 60,000 passengers per year.

**A TRAIN IN FRONT OF BUILDING 143.** At the time of its construction (1942), Building W-143 was the largest government building south of the Pentagon. The building is 820 feet wide by 1,040 feet long. Its concrete frame structure was formed from 110,000 cubic feet of concrete and is supported by 17,700 concrete piles and occupies 48 acres of floor space. The tens of thousands of square feet in W-143 include warehouse and office space as well as packing, shipping, and receiving facilities; the fleet requisition office; a cafeteria; a dispensary; a radio station; and a branch of a local bank.

**THE PUBLIC WORKS CENTER PREPARES FOR USS *MISSISSIPPI* (CGN 40) COMMISSIONING, AUGUST 1978.** The 39th president (and a former naval officer), Jimmy Carter, commissioned the ship. *Mississippi* was powered by a nuclear reactor and was built at nearby Newport News, Virginia.

**A WORKING SAILOR AWAITS THE ARRIVAL OF USS *JOHN F. KENNEDY* (CVA 67) AT PIER 12, OCTOBER 1968.** The *Kennedy* had just been commissioned in September. She left Pier 12 on November 2 for Guantanamo Bay, Cuba, for shakedown training. While at "Gitmo," the ship conducted almost daily general quarters and both day and night flight operations.

**New Bachelor Enlisted Quarters (BEQs) under Construction, November 1971.** The headquarters complex is the center group of buildings at the top of the image.

**USS *Dwight D. Eisenhower* (CVN 69) Moored at the Pier Awaiting Her Commissioning, October 1977.** On October 18, a crowd of thousands at Pier 12 heard the order "Break the commissioning pennant," exactly two years and seven days since her christening.

**Naval Aviation Depot (NADEP) Electricians Rewire an F-14 Cockpit, November 1989.** The aircraft had been engulfed in saltwater while aboard USS *America* (CV 66). The depot celebrated 31 years of mishap-free flying in 1988 when it won the Naval Aviation Command Aviation Safety Award.

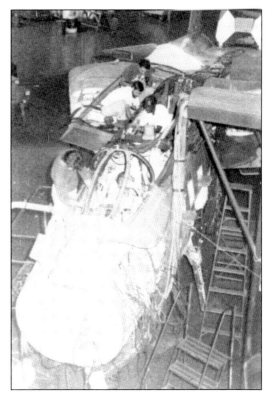

**The Cold War Ends at the Naval Station.** A landmark in the changing relationship between Russia and America was the unprecedented 1989 visit of three Soviet naval vessels to the Naval Base.

**THE NAVAL STATION IN AUGUST 1987, FROM 12,000 FEET.** At this time, about 70,000 vehicles passed through 13 gates of the station each day. In 1987, the base commander, in an annual message, reminded all that the appearance of the base "is an area that demands constant attention . . . men and women deserve an attractive and clean working environment."

**USS ANZIO (CG 68) COMMISSIONING AT THE NAVAL STATION, MAY 1992.** The act of placing a ship in commission marks her entry into active Navy service. At the moment when the commissioning pennant is broken at the masthead, a ship becomes a Navy command in her own right and takes her place alongside the other active ships of the fleet. This ceremony continues a tradition some three centuries old, observed by navies around the world and by our own Navy since 1775.

**PULLING INTO THE NAVAL STATION, MARCH 1991.** Crew members of the nuclear-powered cruiser USS *Mississippi* (CGN 40) line the rails and look out at the Naval Station piers following their return from the Persian Gulf.

**REAR ADM. BYRON "JAKE" TOBIN, COMMANDER NAVAL BASE NORFOLK, APRIL 1991.** Admiral Tobin is observing the waterfront as ships return from Operation Desert Storm. Admiral Tobin was aboard one of the Naval Station's YTB large harbor tugs.

**FIREFIGHTING TRAINING.** A student hose team battles a blaze engulfing a simulated aircraft during an exercise at the Fleet Training Center Firefighting School in 1988. Firefighting courses ensure sailors can control fires at sea, where there is nowhere to turn for help.

**HANDLING A MOORING LINE, MARCH 1991.** Sailors prepare to secure the destroyer tender USS *Yellowstone* (AD 41) to a pier. *Yellowstone* had just returned home from deployment in the Persian Gulf during Operation Desert Storm, March 1991.

**WATER SURVIVAL SKILLS.**
A naval officer practices
survival skills at the Naval
Aviation Survival Training
Facility pool at the Naval
Station in May 1997. At sea,
the lives of the crew rest on
training for emergencies and
on fellow crew members.

**USS WISCONSIN (BB 64) COMES HOME.** Sailors onboard USS *Wisconsin* (BB 64) wait for the ship
to tie up after returning from the Persian Gulf and Operation Desert Storm in 1991. *Wisconsin*
decommissioned in 1992 and became a museum ship in downtown Norfolk in 2001.

**AMERICAN CAPITAL SHIPS AT THE NAVAL STATION PIERS, AUGUST 1992.** From left to right, the carriers are USS *Dwight D. Eisenhower* (CVN 69), USS *Theodore Roosevelt* (CVN 71), USS *America* (CV 66), and USS *George Washington* (CVN 73). *George Washington* had just been commissioned at the Naval Station on July 4.

# *Ten*

# SERVICE TO THE FLEET

At the end of the 20th century the Naval Station's land use plan took stock of the complex's scope and responsibilities: "The 4,600 acre site represents the home for 150 tenant organizations and over 84,000 military personnel." These commands and these people, as has been the case for nearly a century, support the operational readiness of the US Atlantic Fleet. The job includes port services for all ships under naval control in coordination with Atlantic Fleet commands. Activities on the busy waterfront include docking and undocking ships, towing, and firefighting services. Navy personnel deliver potable water, explosives, and bulk fuel to waterfront customers. In addition, the Naval Station assigns berths and anchorages, provides pilots, schedules towing services, and controls harbor movements in the Hampton Roads area.

The station also provides military personnel and their dependents information on what ships are present and the ships' arrival and departure times as well as disseminating weather information.

Along with the military support activities listed above, station managers must address energy and its wise conservation, housing for Navy personnel and their families, food and recreation, communications, physical security, transportation, and the environment.

Entering the 21st century, the station continued its role as a hub of maritime activity on the eastern seaboard and as an integral part of the nation's defenses. The final set of images in this chapter highlight ships that call the Naval Station home.

**USS Arleigh Burke (DDG 51) Under Way in Rough Seas.** USS *Arleigh Burke* is the lead ship of the *Arleigh Burke* class of guided missile destroyers. These ships were the first destroyers in the world equipped with the AEGIS Weapons Systems, which allows the ship to launch,

track, and evade missiles. The ship's namesake, Adm. Arleigh A. Burke, USN, was present at her commissioning in Norfolk on July 4, 1991.

**SERVING AN AIRCRAFT CARRIER.** Fleet oiler USS *Pawcatuck* conducts "underway replenishment" operations with USS *Abraham Lincoln* (CVN 72). *Abraham Lincoln* was built at Newport News, Virginia. She is the Navy's fifth Nimitz-class carrier. These nuclear-powered ships are the largest warships ever built. Underway replenishment is the Navy's term for refueling and resupplying a ship at sea.

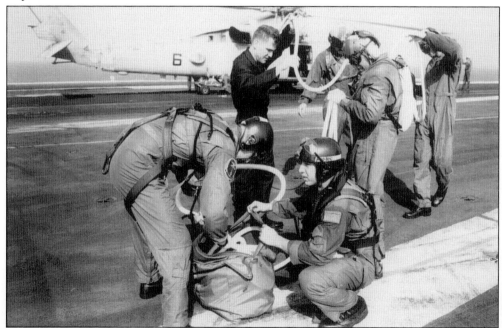

**SAILORS AT WORK ON USS *ABRAHAM LINCOLN* (CVN 72).** Members of Explosive Ordnance Disposal (EOD) Mine Unit 4 pack a rope bag for rigging exercises aboard the carrier. *Abraham Lincoln* and her carrier battle group and air wing helped deliver the opening salvos and air strikes in Operation Iraqi Freedom in 2003.

**USS BARRY (DDG 52).** Destroyers like the *Barry*, an *Arleigh Burke*–class ship, can operate independently or as part of carrier strike groups, surface action groups, amphibious ready groups, and underway replenishment groups.

**SAILORS ABOARD USS HARRY S. TRUMAN (CVN 75).** This picture was taken during the ship's commissioning ceremony at the Naval Station. Like the other Nimitz-class carriers, the *Truman* uses two nuclear reactors to power the ship, which can attain speeds of over 30 knots (34 miles per hour). The ship can carry over 60 aircraft.

**USS NEWPORT NEWS (SSN 750).** This nuclear-powered *Los Angeles*–class submarine was built at Newport News, Virginia, the city for which she was named. She was the third ship to be named for the famous maritime city on the James River.

**USS THEODORE ROOSEVELT (CVN 71) HEADS DOWN THE ELIZABETH RIVER TO SEA.** An American carrier is often described as a "city at sea." *Theodore Roosevelt* carries a ship's company of over 3,500 sailors with another 1,500 serving in the air wing.

**THE TIMELESS SCENE OF A NAVY HOMECOMING.** A sailor from USS *America* (CV 66) is reunited with a loved one following the ship's deployment to the Persian Gulf during Operation Desert Storm in 1991.

# DISCOVER THOUSANDS OF LOCAL HISTORY BOOKS
## FEATURING MILLIONS OF VINTAGE IMAGES

Arcadia Publishing, the leading local history publisher in the United States, is committed to making history accessible and meaningful through publishing books that celebrate and preserve the heritage of America's people and places.

Find more books like this at
## www.arcadiapublishing.com

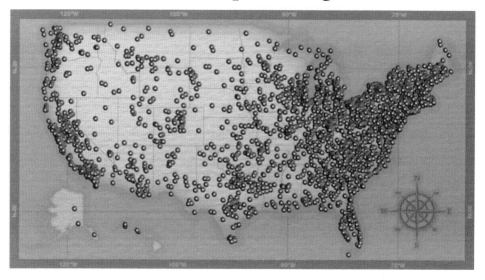

Search for your hometown history, your old stomping grounds, and even your favorite sports team.

Consistent with our mission to preserve history on a local level, this book was printed in South Carolina on American-made paper and manufactured entirely in the United States. Products carrying the accredited Forest Stewardship Council (FSC) label are printed on 100 percent FSC-certified paper.

**MADE IN THE**